Copyright © 2011 XAMonline, Inc.
All rights reserved. No part of the material protected by this copyright notice may be reproduced or utilized in any form or by any means, electronic or mechanical, including photocopying, recording or by any information storage and retrievable system, without written permission from the copyright holder.

To obtain permission(s) to use the material from this work for any purpose including workshops or seminars, please submit a written request to:

XAMonline, Inc.
25 First Street, Suite 106
Cambridge, MA 02141
Toll Free: 1-800-509-4128
Email: info@xamonline.com
Web: www.xamonline.com
Fax: 1-617-583-5552

Library of Congress Cataloging-in-Publication Data

Wynne, Sharon A.
    ILTS Mathematics 115 Practice Test 1: Teacher Certification /
Sharon A. Wynne. -1$^{st}$ ed.
    ISBN: 978-1-60787-203-0
    1. ILTS Mathematics 115 Practice Test 1
    2. Study Guides       3. ILTS      4. Teachers' Certification & Licensure
    5. Careers

**Disclaimer:**
The opinions expressed in this publication are the sole works of XAMonline and were created independently from the National Education Association, Educational Testing Service, or any State Department of Education, National Evaluation Systems or other testing affiliates.

Between the time of publication and printing, state specific standards as well as testing formats and website information may change that is not included in part or in whole within this product. Sample test questions are developed by XAMonline and reflect similar content as on real tests; however, they are not former tests. XAMonline assembles content that aligns with state standards but makes no claims nor guarantees teacher candidates a passing score. Numerical scores are determined by testing companies such as NES or ETS and then are compared with individual state standards. A passing score varies from state to state.

**Printed in the United States of America**     œ-1
ILTS Mathematics 115 Practice Test 1
ISBN: 978-1-60787-203-0

# Mathematics
## Pre-Test Sample Questions

1. Which of the following sets is closed under division?
   *(Average)*

   I. {½, 1, 2, 4}
   II. {-1, 1}
   III. {-1, 0, 1}

   A. I only

   B. II only

   C. III only

   D. I and II

2. Which of the following is always composite if *x* is odd, *y* is even, and both *x* and *y* are greater than or equal to 2?
   *(Average)*

   A. $x + y$

   B. $3x + 2y$

   C. $5xy$

   D. $5x + 3y$

3. Find the LCM of 27, 90, and 84.
   *(Easy)*

   A. 90

   B. 3,780

   C. 204,120

   D. 1,260

4. Solve for $x$: $18 = 4 + |2x|$
   *(Rigorous)*

   A. $\{-11, 7\}$

   B. $\{-7, 0, 7\}$

   C. $\{-7, 7\}$

   D. $\{-11, 11\}$

5. Which graph represents the solution set for $x^2 - 5x > -6$?
   *(Average)*

   A. ←―⊕―――⊕―→
         -2  0  2

   B. ←⊕―――――⊕→
        -3   0   3

   C. ←―⊕―――⊕――→
         -2  0  2

   D. ←――――――⊕⊕→
              0  2 3

**MATHEMATICS**

6. What is the equation of the graph shown below?
   *(Easy)*

   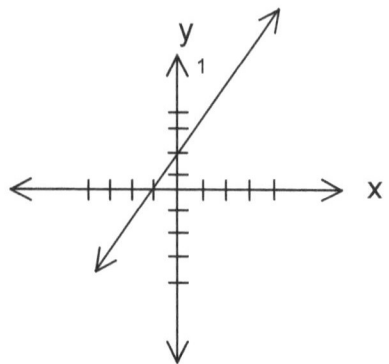

   A. $2x + y = 2$

   B. $2x - y = -2$

   C. $2x - y = 2$

   D. $2x + y = -2$

7. Find the length of the major axis of $x^2 + 9y^2 = 36$.
   *(Rigorous)*

   A. 4

   B. 6

   C. 12

   D. 8

8. Which equation represents a circle with a diameter whose endpoints are $(0,7)$ and $(0,3)$?
   *(Rigorous)*

   A. $x^2 + y^2 + 21 = 0$

   B. $x^2 + y^2 - 10y + 21 = 0$

   C. $x^2 + y^2 - 10y + 9 = 0$

   D. $x^2 - y^2 - 10y + 9 = 0$

9. Given that *M* is a mass, *V* is a velocity, *A* is an acceleration, and *T* is a time, what type of unit corresponds to the overall expression $\frac{AMT}{V}$?
   *(Average)*

   A. Mass

   B. Time

   C. Velocity

   D. Acceleration

10. The term "cubic feet" indicates which kind of measurement?
    *(Average)*

    A. Volume

    B. Mass

    C. Length

    D. Distance

11. Which word best describes a set of measured values that are all very similar but that all deviate significantly from the expected result?
    *(Easy)*

    A. Perfect

    B. Precise

    C. Accurate

    D. Appropriate

12. When you begin by assuming the conclusion of a theorem is false, then show that through a sequence of logically correct steps you contradict an accepted fact, this is known as
    *(Easy)*

    A. Inductive reasoning

    B. Direct proof

    C. Indirect proof

    D. Exhaustive proof

13. Compute the area of the shaded region, given a radius of 5 meters. Point O is the center.
    *(Rigorous)*

    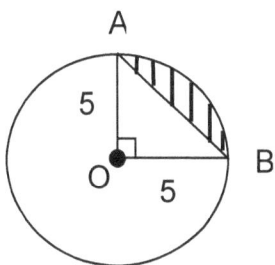

    A. 7.13 cm²

    B. 7.13 m²

    C. 78.5 m²

    D. 19.63 m²

14. Given a 30-by-60-meter garden with a circular fountain with a 5 meter radius, calculate the area of the portion of the garden not occupied by the fountain.
    *(Average)*

    A. 1,721 m²

    B. 1,879 m²

    C. 2,585 m²

    D. 1,015 m²

**MATHEMATICS**

15. If the area of the base of a cone is tripled, the volume will be
    (Rigorous)

    A. The same as the original

    B. 9 times the original

    C. 3 times the original

    D. $3\pi$ times the original

16. Find the area of the figure pictured below.
    (Rigorous)

    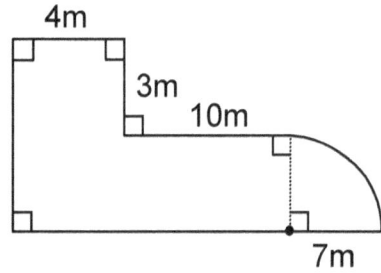

    A. 136.47 m²

    B. 148.48 m²

    C. 293.86 m²

    D. 178.47 m²

17. Determine the measures of angles A and B.
    (Average)

    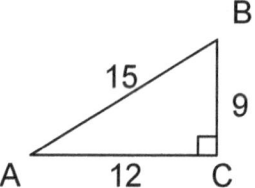

    A. A = 30°, B = 60°

    B. A = 60°, B = 30°

    C. A = 53°, B = 37°

    D. A = 37°, B = 53°

18. Which expression is not identical to sin x?
    (Average)

    A. $\sqrt{1 - \cos^2 x}$

    B. $\tan x \cos x$

    C. $1 / \csc x$

    D. $1 / \sec x$

19. The following is a Pythagorean identity:
    (Easy)

    A. $\sin^2 \theta - \cos^2 \theta = 1$

    B. $\sin^2 \theta + \cos^2 \theta = 1$

    C. $\cos^2 \theta - \sin^2 \theta = 1$

    D. $\cos^2 \theta + \tan^2 \theta = 1$

MATHEMATICS

20. Which expression is equivalent to $1-\sin^2 x$?
    *(Rigorous)*

    A. $1-\cos^2 x$

    B. $1+\cos^2 x$

    C. $1/\sec x$

    D. $1/\sec^2 x$

21. State the domain of the function $f(x)=\dfrac{3x-6}{x^2-25}$
    *(Average)*

    A. $x \neq 2$

    B. $x \neq 5, -5$

    C. $x \neq 2, -2$

    D. $x \neq 5$

22. Which of the following is a factor of the expression $9x^2 + 6x - 35$?
    *(Rigorous)*

    A. $3x - 5$

    B. $3x - 7$

    C. $x + 3$

    D. $x - 2$

23. Which of the following is a factor of $6+48m^3$?
    *(Rigorous)*

    A. $(1 + 2m)$

    B. $(1 - 8m)$

    C. $(1 + m - 2m)$

    D. $(1 - m + 2m)$

24. Solve for $x$ by factoring $2x^2 - 3x - 2 = 0$.
    *(Average)*

    A. $x = (-1, 2)$

    B. $x = (0.5, -2)$

    C. $x = (-0.5, 2)$

    D. $x = (1, -2)$

25. Which of the following is equivalent to $\sqrt[b]{x^a}$?
    *(Easy)*

    A. $x^{a/b}$

    B. $x^{b/a}$

    C. $a^{x/b}$

    D. $b^{x/a}$

**MATHEMATICS**

26. Which graph represents the equation of $y = x^2 + 3x$?
    *(Average)*

    A.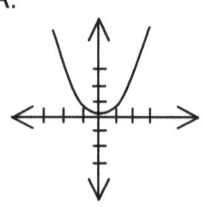
    B.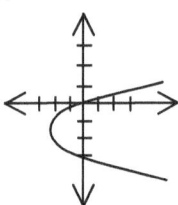

    C.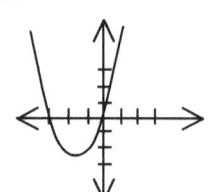
    D.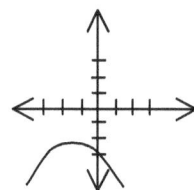

27. Given $f(x) = 3x - 2$ and $g(x) = x^2$, determine $g(f(x))$.
    *(Average)*

    A. $3x^2 - 2$

    B. $9x^2 + 4$

    C. $9x^2 - 12x + 4$

    D. $3x^3 - 2$

28. L'Hospital's rule provides a method to evaluate which of the following?
    *(Easy)*

    A. Limit of a function

    B. Derivative of a function

    C. Sum of an arithmetic series

    D. Sum of a geometric series

29. Find the first derivative of the function: $f(x) = x^3 - 6x^2 + 5x + 4$
    *(Rigorous)*

    A. $3x^3 - 12x^2 + 5x$

    B. $3x^2 - 12x - 5$

    C. $3x^2 - 12x + 9$

    D. $3x^2 - 12x + 5$

30. Find the absolute maximum obtained by the function $y = 2x^2 + 3x$ on the interval $x = 0$ to $x = 3$.
    *(Rigorous)*

    A. $-3/4$

    B. $-4/3$

    C. 0

    D. 27

31. How does the function $y = x^3 + x^2 + 4$ behave from $x = 1$ to $x = 3$?
    *(Average)*

    A. Increasing, then decreasing

    B. Increasing

    C. Decreasing

    D. Neither increasing nor decreasing

32. Find the area under the function $y = x^2 + 4$ from $x = 3$ to $x = 6$.
    *(Average)*

    A. 75

    B. 21

    C. 96

    D. 57

33. If the velocity of a body is given by $v = 16 - t^2$, find the distance traveled from $t = 0$ until the body comes to a complete stop.
    *(Average)*

    A. 16

    B. 43

    C. 48

    D. 64

34. Find the antiderivative for $4x^3 - 2x + 6 = y$.
    *(Rigorous)*

    A. $x^4 - x^2 + 6x + C$

    B. $x^4 - 2/3x^3 + 6x + C$

    C. $12x^2 - 2 + C$

    D. $4/3x^4 - x^2 + 6x + C$

35. Compute the standard deviation for the following set of temperatures:

    (37, 38, 35, 37, 38, 40, 36, 39)

    *(Rigorous)*

    A. 37.5

    B. 1.5

    C. 0.5

    D. 2.5

36. Compute the median for the following data set:

    {12, 19, 13, 16, 17, 14}

    *(Easy)*

    A. 14.5

    B. 15.17

    C. 15

    D. 16

**MATHEMATICS**

37. Half the students in a class scored 80% on an exam, most of the rest scored 85% except for one student who scored 10%. Which would be the best measure of central tendency for the test scores?
*(Rigorous)*

    A. Mean

    B. Median

    C. Mode

    D. Either the median or the mode because they are equal

38. If the correlation between two variables is given as zero, the association between the two variables is
*(Rigorous)*

    A. Negative linear

    B. Positive linear

    C. Quadratic

    D. Random

39. Which of the following is not a valid method of collecting statistical data?
*(Average)*

    A. Random sampling

    B. Systematic sampling

    C. Cluster sampling

    D. Cylindrical sampling

40. A jar contains 3 red marbles, 5 white marbles, 1 green marble, and 15 blue marbles. If one marble is picked at random from the jar, what is the probability that it will be red?
*(Easy)*

    A. 1/3

    B. 1/8

    C. 3/8

    D. 1/24

41. A die is rolled several times. What is the probability that a 3 will not appear before the third roll of the die?
*(Rigorous)*

    A. 1/3

    B. 25/216

    C. 25/36

    D. 1/216

**MATHEMATICS**

42. A baseball team has a 60% chance of winning any particular game in a seven-game series. What is the probability that it will win the series by winning games six and seven?
*(Rigorous)*

   A. 8.3%

   B. 36%

   C. 50%

   D. 60%

43. The scalar multiplication of the number 3 with the matrix $\begin{pmatrix} 2 & 1 \\ 3 & 5 \end{pmatrix}$ yields:
*(Easy)*

   A. 33

   B. $\begin{pmatrix} 6 & 1 \\ 9 & 5 \end{pmatrix}$

   C. $\begin{pmatrix} 2 & 3 \\ 3 & 15 \end{pmatrix}$

   D. $\begin{pmatrix} 6 & 3 \\ 9 & 15 \end{pmatrix}$

44. Evaluate the following matrix product:

   $\begin{pmatrix} 2 & 1 & 3 \\ 2 & 2 & 4 \end{pmatrix} \begin{pmatrix} 6 & 5 \\ 2 & 1 \\ 2 & 7 \end{pmatrix}$

   *(Rigorous)*

   A. $\begin{pmatrix} 20 & 32 & 24 \\ 24 & 40 & 48 \end{pmatrix}$

   B. $\begin{pmatrix} 20 & 32 \\ 40 & 24 \\ 24 & 48 \end{pmatrix}$

   C. 116

   D. $\begin{pmatrix} 20 & 32 \\ 24 & 40 \end{pmatrix}$

45. Solve the following matrix equation:

   $3x + \begin{pmatrix} 1 & 5 & 2 \\ 0 & 6 & 9 \end{pmatrix} = \begin{pmatrix} 7 & 17 & 5 \\ 3 & 9 & 9 \end{pmatrix}$

   *(Average)*

   A. $\begin{pmatrix} 2 & 4 & 1 \\ 1 & 1 & 0 \end{pmatrix}$

   B. 2

   C. $\begin{pmatrix} 8 & 23 & 7 \\ 3 & 15 & 18 \end{pmatrix}$

   D. $\begin{pmatrix} 9 \\ 2 \end{pmatrix}$

**MATHEMATICS**

46. Find the value of the determinant of the matrix.

$$\begin{vmatrix} 2 & 1 & -1 \\ 4 & -1 & 4 \\ 0 & -3 & 2 \end{vmatrix}$$

*(Average)*

A. 0

B. 23

C. 24

D. 40

47. Which of the following is a recursive definition of the sequence $\{1, 2, 2, 4, 8, 32, \ldots\}$?
*(Average)*

A. $N_i = 2N_{i-1}$

B. $N_i = 2N_{i-2}$

C. $N_i = N_{i-1}^2$

D. $N_i = N_{i-1} N_{i-2}$

48. What is the sum of the first 20 terms of the geometric sequence:

(2, 4, 8, 16, 32…)?

*(Rigorous)*

A. 2,097,150

B. 1,048,575

C. 524,288

D. 1,048,576

49. Find the sum of the first one hundred terms in the progression.

(−6, −2, 2…)

*(Rigorous)*

A. 19,200

B. 19,400

C. −604

D. 604

**MATHEMATICS**

50. If an initial deposit of $1,000 is made to a savings account with a continuously compounded annual interest rate of 5%, how much money is in the account after 4.5 years? *(Average)*

    A. $1,200.00

    B. $1,225.00

    C. $1,245.52

    D. $1,252.32

**MATHEMATICS**

# Mathematics
## Pre-Test Sample Questions with Rationales

1. Which of the following sets is closed under division?
   *(Average)*

   I.   {½, 1, 2, 4}
   II.  {-1, 1}
   III. {-1, 0, 1}

   A. I only

   B. II only

   C. III only

   D. I and II

**Answer: B. II only**

I is not closed because $\frac{4}{.5} = 8$ and 8 is not in the set. III is not closed because $\frac{1}{0}$ is undefined. II is closed because $\frac{-1}{1} = -1, \frac{1}{-1} = -1, \frac{1}{1} = 1, \frac{-1}{-1} = 1$ are all in the set.

2. Which of the following is always composite if *x* is odd, *y* is even, and both *x* and *y* are greater than or equal to 2?
   *(Average)*

   A. $x + y$

   B. $3x + 2y$

   C. $5xy$

   D. $5x + 3y$

**Answer: C.** $5xy$

A composite number is a number that is not prime. The prime number sequence begins 2,3,5,7,11,13,17,.... To determine which of the expressions is always composite, experiment with different values of *x* and *y*, such as *x* = 3 and *y* = 2, or *x* = 5 and *y* = 2. It turns out that 5*xy* will always be an even number and, therefore, composite if *y* = 2.

**MATHEMATICS**

3. **Find the LCM of 27, 90, and 84.**
   *(Easy)*

   A. 90

   B. 3,780

   C. 204,120

   D. 1,260

**Answer: B. 3,780**
To find the LCM of the above numbers, factor each into its prime factors and multiply each common factor the maximum number of times it occurs. Thus, 27 = 3x3x3; 90 = 2x3x3x5; 84 = 2x2x3x7; LCM = 2x2x3x3x3x5x7 = 3,780.

4. **Solve for $x$: $18 = 4 + |2x|$**
   *(Rigorous)*

   A. $\{-11, 7\}$

   B. $\{-7, 0, 7\}$

   C. $\{-7, 7\}$

   D. $\{-11, 11\}$

**Answer: C. $\{-7, 7\}$**
Using the definition of absolute value, two equations are possible: 18 = 4 + 2x or 18 = 4 − 2x. Solving for $x$ yields $x$ = 7 or $x$ = −7.

**MATHEMATICS**

5. Which graph represents the solution set for $x^2 - 5x > -6$?
   *(Average)*

   A. ◄─○───○──► 
      −2  0  2

   B. ◄○──────○─► 
      −3  0   3

   C. ◄─○───○──► 
      −2  0  2

   D. ◄──────○○─► 
             0  2 3

   **Answer: D.** ◄──────○○─►
                        0  2 3

   Rewriting the inequality yields $x^2 - 5x + 6 > 0$. Factoring yields $(x - 2)(x - 3) > 0$. The two cut-off points on the number line are now at $x = 2$ and $x = 3$. Choosing a random number in each of the three parts of the number line, test them to see if they produce a true statement. If $x = 0$ or $x = 4$, $(x - 2)(x - 3) > 0$ is true. If $x = 2.5$, $(x - 2)(x - 3) > 0$ is false. Therefore, the solution set is all numbers smaller than 2 or greater than 3.

**MATHEMATICS**

6. What is the equation of the graph shown below?
   *(Easy)*

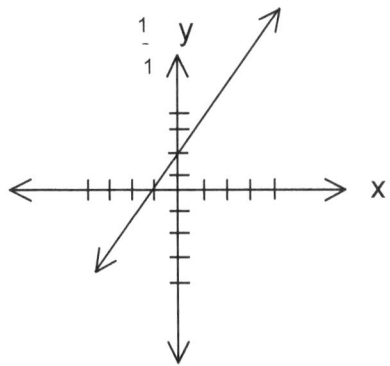

   A.   $2x + y = 2$

   B.   $2x - y = -2$

   C.   $2x - y = 2$

   D.   $2x + y = -2$

**Answer: B.** $2x - y = -2$

By observation, we see that the graph has a *y*-intercept of 2 and a slope of 2/1 = 2. Therefore, its equation is *y* = *mx* + *b* = 2*x* + 2. Rearranging the terms yields 2*x* – *y* = –2.


7. Find the length of the major axis of $x^2 + 9y^2 = 36$.
   *(Rigorous)*

   A.   4

   B.   6

   C.   12

   D.   8

**Answer: C. 12**

Dividing by 36 yields $\frac{x^2}{36} + \frac{y^2}{4} = 1$, which implies that the ellipse intersects the *x*-axis at 6 and –6. Therefore, the length of the major axis is 12.


**MATHEMATICS**

8. Which equation represents a circle with a diameter whose endpoints are $(0,7)$ and $(0,3)$?

*(Rigorous)*

    A. $\quad x^2 + y^2 + 21 = 0$

    B. $\quad x^2 + y^2 - 10y + 21 = 0$

    C. $\quad x^2 + y^2 - 10y + 9 = 0$

    D. $\quad x^2 - y^2 - 10y + 9 = 0$

**Answer: B.** $x^2 + y^2 - 10y + 21 = 0$

With a diameter going from (0,7) to (0,3), the diameter of the circle must be 4, the radius must be 2 and the center of the circle must be at (0,5). Using the standard form for the equation of a circle, we get $(x-0)^2 + (y-5)^2 = 2^2$. Expanding yields $x^2 + y^2 - 10y + 21 = 0$.

**MATHEMATICS**

9. Given that *M* is a mass, *V* is a velocity, *A* is an acceleration, and *T* is a time, what type of unit corresponds to the overall expression $\frac{AMT}{V}$?
   *(Average)*

   A.  Mass

   B.  Time

   C.  Velocity

   D.  Acceleration

**Answer: A. Mass**
Use unit analysis to find the simplest expression for the units associated with the expression. Choose any unit system: for example, the metric system.

$$\frac{AMT}{V} = \frac{\left(\frac{m}{s^2}\right)(kg)(s)}{\left(\frac{m}{s}\right)}$$

Simplify the units in the expression

$$\frac{AMT}{V} = \frac{\left(\frac{m}{s}\right)(kg)}{\left(\frac{m}{s}\right)} = kg$$

Kilograms are a unit of mass, and thus the correct answer is A.

10. The term "cubic feet" indicates which kind of measurement?
    *(Average)*

    A.  Volume

    B.  Mass

    C.  Length

    D.  Distance

**Answer: A. Volume**
The word *cubic* indicates that this is a term describing volume.

**MATHEMATICS**

11. Which word best describes a set of measured values that are all very similar but that all deviate significantly from the expected result?
    *(Easy)*

    A.  Perfect

    B.  Precise

    C.  Accurate

    D.  Appropriate

**Answer: B. Precise**
A set of measurements that are close to the same value is *precise*. Measurements that are close to the actual (or expected) value are *accurate*. In this case, the set of measurements described in the question are best summarized as precise.

12. When you begin by assuming the conclusion of a theorem is false, then show that through a sequence of logically correct steps you contradict an accepted fact, this is known as
    *(Easy)*

    A.  Inductive reasoning

    B.  Direct proof

    C.  Indirect proof

    D.  Exhaustive proof

**Answer: C. Indirect proof**
By definition, this describes the procedure of an indirect proof.

**MATHEMATICS**

13. Compute the area of the shaded region, given a radius of 5 meters. Point O is the center.
    *(Rigorous)*

    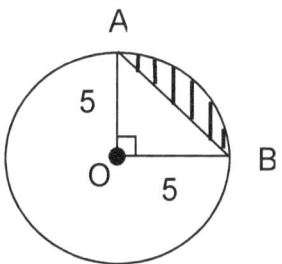

    A. 7.13 cm²

    B. 7.13 m²

    C. 78.5 m²

    D. 19.63 m²

**Answer: B. 7.13 m²**

The area of triangle AOB is .5(5)(5) = 12.5 square meters. Since $\frac{90}{360} = .25$, the area of sector AOB (the pie-shaped piece) is approximately $.25(\pi)5^2$ = 19.63. Subtracting the triangle area from the sector area to get the area of segment AB yields approximately 19.63 – 12.5 = 7.13 square meters.

14. Given a 30-by-60-meter garden with a circular fountain with a 5 meter radius, calculate the area of the portion of the garden not occupied by the fountain.
    *(Average)*

    A. 1,721 m²

    B. 1,879 m²

    C. 2,585 m²

    D. 1,015 m²

**Answer: A. 1,721 m²**

Find the area of the garden and then subtract the area of the fountain: 30(60) – $\pi(5)^2$ or approximately 1,721 square meters.

**MATHEMATICS**

15. **If the area of the base of a cone is tripled, the volume will be**
    *(Rigorous)*

    A.  The same as the original

    B.  9 times the original

    C.  3 times the original

    D.  3π times the original

**Answer: C. 3 times the original**

The formula for the volume of a cone is $V = \frac{1}{3}Bh$, where B is the area of the circular base and h is the height. If the area of the base is tripled, the volume becomes $V = \frac{1}{3}(3B)h = Bh$, or three times the original area.

16. **Find the area of the figure pictured below.**
    *(Rigorous)*

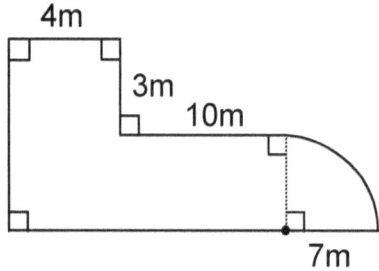

   A.  136.47 m²

   B.  148.48 m²

   C.  293.86 m²

   D.  178.47 m²

**Answer: B. 148.48 m²**
Divide the figure into 2 rectangles and one-quarter circle. The tall rectangle on the left has dimensions of 10 by 4 and an area of 40. The rectangle in the center has dimensions 7 by 10 and area 70. The quarter circle has area $.25(\pi)7^2$ = 38.48. The total area is therefore approximately 148.48.

**MATHEMATICS**

17. Determine the measures of angles A and B.
    *(Average)*

    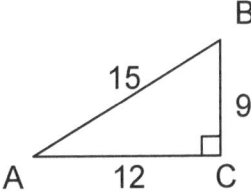

    A.   A = 30°, B = 60°

    B.   A = 60°, B = 30°

    C.   A = 53°, B = 37°

    D.   A = 37°, B = 53°

**Answer: D. A = 37°, B = 53°**
It is the case that tan A = 9/12 = .75 and $\tan^{-1}$ .75 = 37 degrees. Since angle B is complementary to angle A, the measure of angle B is therefore 53 degrees.

18. Which expression is not identical to sin x?
    *(Average)*

    A.   $\sqrt{1 - \cos^2 x}$

    B.   $\tan x \cos x$

    C.   $1 / \csc x$

    D.   $1 / \sec x$

**Answer: D.** $1 / \sec x$
Using the basic definitions of the trigonometric functions and the Pythagorean identity, it can be seen that the first three options are all identical to sin *x*. sec *x* = 1/cos *x* is not the same as sin *x*.

**MATHEMATICS**

19. The following is a Pythagorean identity:
    *(Easy)*

    A.  $\sin^2 \theta - \cos^2 \theta = 1$

    B.  $\sin^2 \theta + \cos^2 \theta = 1$

    C.  $\cos^2 \theta - \sin^2 \theta = 1$

    D.  $\cos^2 \theta + \tan^2 \theta = 1$

**Answer: B.** $\sin^2 \theta + \cos^2 \theta = 1$

The Pythagorean identity $\sin^2 \theta + \cos^2 \theta = 1$ is derived from the definitions of the sine and cosine functions and Pythagorean theorem of geometry.

20. Which expression is equivalent to $1 - \sin^2 x$ ?
    *(Rigorous)*

    A.  $1 - \cos^2 x$

    B.  $1 + \cos^2 x$

    C.  $1/\sec x$

    D.  $1/\sec^2 x$

**Answer: D.** $1/\sec^2 x$

Using the Pythagorean Identity, it is apparent that $\sin^2 x + \cos^2 x = 1$.
Thus, $1 - \sin^2 x = \cos^2 x$, which by definition is equal to $1/\sec^2 x$.

**MATHEMATICS**

21. **State the domain of the function** $f(x) = \dfrac{3x-6}{x^2-25}$

    *(Average)*

    A.   $x \neq 2$

    B.   $x \neq 5, -5$

    C.   $x \neq 2, -2$

    D.   $x \neq 5$

**Answer: B.** $x \neq 5, -5$

The values 5 and –5 must be omitted from the domain because if x took on either of those values, the denominator of the fraction would have a value of 0. Therefore the fraction would be undefined. Thus, the domain of the function is all real numbers x not equal to 5 and –5.

22. **Which of the following is a factor of the expression** $9x^2 + 6x - 35$?
    *(Rigorous)*

    A.   3x – 5

    B.   3x – 7

    C.   x + 3

    D.   x – 2

**Answer: A. 3x – 5**

Recognize that the given expression can be written as the sum of two squares and use the formula $a^2 - b^2 = (a+b)(a-b)$.

$$9x^2 + 6x - 35 = (3x+1)^2 - 36 = (3x+1+6)(3x+1-6) = (3x+7)(3x-5)$$

**MATHEMATICS**

23. **Which of the following is a factor of** $6+48m^3$ **?**
    *(Rigorous)*

    A.  $(1 + 2m)$

    B.  $(1 - 8m)$

    C.  $(1 + m - 2m)$

    D.  $(1 - m + 2m)$

**Answer: A. (1 + 2m)**
Removing the common factor of 6 and then factoring the sum of two cubes yields
$6 + 48m^3 = 6(1 + 8m^3) = 6(1 + 2m)(1^2 - 2m + (2m)^2)$.

24. **Solve for *x* by factoring** $2x^2 - 3x - 2 = 0$.
    *(Average)*

    A.  x = (–1, 2)

    B.  x = (0.5, –2)

    C.  x = (–0.5, 2)

    D.  x = (1, –2)

**Answer: C. x = (–0.5, 2)**
Because $2x^2 - 3x - 2 = 2x^2 - 4x + x - 2 = 2x(x-2) + (x-2) = (2x+1)(x-2) = 0$,
then x = –0.5 or 2.

**MATHEMATICS**

25. Which of the following is equivalent to $\sqrt[b]{x^a}$ ?
    *(Easy)*

    A. $x^{a/b}$

    B. $x^{b/a}$

    C. $a^{x/b}$

    D. $b^{x/a}$

**Answer: A.** $x^{a/b}$

The $b^{th}$ root, expressed in the form $\sqrt[b]{\phantom{x}}$, can also be written as an exponential, $1/b$. Writing the expression in this form, $(x^a)^{1/b}$, and then multiplying exponents yields $x^{a/b}$.

**MATHEMATICS**

26. Which graph represents the equation of $y = x^2 + 3x$?
    *(Average)*

    A. 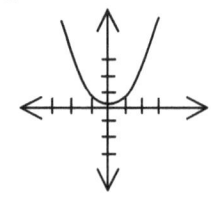　　B.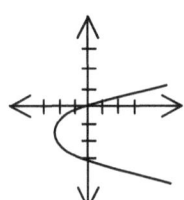

    C. 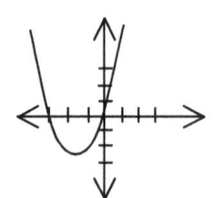　　D.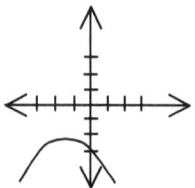

**Answer: C.**

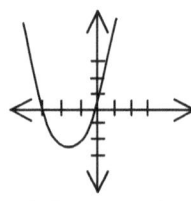

Choice B is not the graph of a function. Choice D is the graph of a parabola where the coefficient of $x^2$ is negative. Choice A appears to be the graph of $y = x^2$. To find the x-intercepts of $y = x^2 + 3x$, set $y = 0$ and solve for $x$: $0 = x^2 + 3x = x(x + 3)$ to get $x = 0$ or $x = -3$. Therefore, the graph of the function intersects the x-axis at $x = 0$ and $x = -3$.

**MATHEMATICS**

27. Given $f(x) = 3x - 2$ and $g(x) = x^2$, determine $g(f(x))$.
    *(Average)*

    A. $3x^2 - 2$

    B. $9x^2 + 4$

    C. $9x^2 - 12x + 4$

    D. $3x^3 - 2$

**Answer: C.** $9x^2 - 12x + 4$
The composite function $g(f(x))$ is $(3x - 2)^2 = 9x^2 - 12x + 4$.

28. L'Hospital's rule provides a method to evaluate which of the following?
    *(Easy)*

    A. Limit of a function

    B. Derivative of a function

    C. Sum of an arithmetic series

    D. Sum of a geometric series

**Answer: A. Limit of a function**
L'Hospital's rule is used to find the limit of a function by taking the derivatives of the numerator and denominator. Since the primary purpose of the rule is to find the limit, A is the correct answer.

**MATHEMATICS**

29. Find the first derivative of the function: $f(x) = x^3 - 6x^2 + 5x + 4$
    *(Rigorous)*

    A.     $3x^3 - 12x^2 + 5x$

    B.     $3x^2 - 12x - 5$

    C.     $3x^2 - 12x + 9$

    D.     $3x^2 - 12x + 5$

**Answer: D.** $3x^2 - 12x + 5$
Use the Power Rule for polynomial differentiation: if $y = ax^n$, then $y' = nax^{n-1}$. Apply this rule to each term in the polynomial to yield the result in answer D.

30. Find the absolute maximum obtained by the function $y = 2x^2 + 3x$ on the interval $x = 0$ to $x = 3$.
    *(Rigorous)*

    A.     −3/4

    B.     −4/3

    C.     0

    D.     27

**Answer: D. 27**
Find the critical point at $x = -.75$ as done in #44. Since the critical point is not in the interval from $x = 0$ to $x = 3$, simply find the values of the function at the endpoints. The endpoints are $x = 0$, $y = 0$, and $x = 3$, $y = 27$. Therefore, 27 is the absolute maximum on the given interval.

**MATHEMATICS**

31. **How does the function $y = x^3 + x^2 + 4$ behave from $x = 1$ to $x = 3$?**
    *(Average)*

    A.  Increasing, then decreasing

    B.  Increasing

    C.  Decreasing

    D.  Neither increasing nor decreasing

**Answer: B. Increasing**
To find critical points, take the derivative of the function and set it equal to 0, and solve for x.

$$f'(x) = 3x^2 + 2x = x(3x + 2)$$

The critical points are at $x = 0$ and $x = -2/3$. Neither of these is on the interval from $x = 1$ to $x = 3$. Test the endpoints: at $x = 1$, $y = 6$, and at $x = 3$, $y = 38$. Since the derivative is positive for all values of x from $x = 1$ to $x = 3$, the curve is increasing on the entire interval.

32. **Find the area under the function $y = x^2 + 4$ from $x = 3$ to $x = 6$.**
    *(Average)*

    A.  75

    B.  21

    C.  96

    D.  57

**Answer: A. 75**
To find the area, set up the definite integral: $\int_{3}^{6}(x^2 + 4)dx = (\frac{x^3}{3} + 4x)$. Evaluate the expression at $x = 6$ and $x = 3$, and then subtract to get $(72 + 24) - (9 + 12) = 75$.

**MATHEMATICS**

33. **If the velocity of a body is given by $v = 16 - t^2$, find the distance traveled from $t = 0$ until the body comes to a complete stop.**
   *(Average)*

   A.  16

   B.  43

   C.  48

   D.  64

**Answer: B. 43**

Recall that the derivative of the distance function is the velocity function. Conversely, the integral of the velocity function is the distance function. To find the time needed for the body to come to a stop when $v = 0$, solve for $t$:
$v = 16 - t^2 = 0$. The result is t = 4 seconds. The distance function is $s = 16t - \frac{t^3}{3}$.
At t = 4, s = 64 − 64/3 or approximately 43 units.

34. **Find the antiderivative for $4x^3 - 2x + 6 = y$.**
   *(Rigorous)*

   A.  $x^4 - x^2 + 6x + C$

   B.  $x^4 - 2/3x^3 + 6x + C$

   C.  $12x^2 - 2 + C$

   D.  $4/3x^4 - x^2 + 6x + C$

**Answer: A.** $x^4 - x^2 + 6x + C$

Use the rule for polynomial integration: given $ax^n$, the antiderivative is $\frac{ax^{n+1}}{n+1}$.
Apply this rule to each term in the polynomial to get the result in answer A.

**MATHEMATICS**

35. Compute the standard deviation for the following set of temperatures:

    (37, 38, 35, 37, 38, 40, 36, 39)

    *(Rigorous)*

    A.   37.5

    B.   1.5

    C.   0.5

    D.   2.5

**Answer: B. 1.5**
First find the mean: 300/8 = 37.5. Then, using the formula for standard deviation yields

$$\sqrt{\frac{2(37.5-37)^2 + 2(37.5-38)^2 + (37.5-35)^2 + (37.5-40)^2 + (37.5-36)^2 + (37.5-39)^2}{8}}$$

This expression has a value of 1.5.

36. Compute the median for the following data set:

    {12, 19, 13, 16, 17, 14}

    *(Easy)*

    A.   14.5

    B.   15.17

    C.   15

    D.   16

**Answer: C. 15**
Arrange the data in ascending order: 12, 13, 14, 16, 17, 19. The median is the middle value in a list with an odd number of entries. When there is an even number of entries, the median is the mean of the two center entries. Here the average of 14 and 16 is 15.

**MATHEMATICS**

37. Half the students in a class scored 80% on an exam, most of the rest scored 85% except for one student who scored 10%. Which would be the best measure of central tendency for the test scores?
*(Rigorous)*

   A.   Mean

   B.   Median

   C.   Mode

   D.   Either the median or the mode because they are equal

**Answer: B. Median**
In this set of data, the median is the most representative measure of central tendency because the median is independent of extreme values. Because of the 10% outlier, the mean (average) would be disproportionately skewed. In this data set, it is true that the median and the mode (number which occurs most often) are the same, but the median remains the best choice because of its special properties.

38. If the correlation between two variables is given as zero, the association between the two variables is
*(Rigorous)*

   A.   Negative linear

   B.   Positive linear

   C.   Quadratic

   D.   Random

**Answer: D. Random**
A correlation of 1 indicates a perfect positive linear association, a correlation of –1 indicates a perfect negative linear association, and a correlation of zero indicates a random relationship between the variables.

**MATHEMATICS**

39. Which of the following is not a valid method of collecting statistical data?
    *(Average)*

    A.  Random sampling

    B.  Systematic sampling

    C.  Cluster sampling

    D.  Cylindrical sampling

**Answer: D. Cylindrical sampling**
There is no such method as cylindrical sampling.


40. A jar contains 3 red marbles, 5 white marbles, 1 green marble, and 15 blue marbles. If one marble is picked at random from the jar, what is the probability that it will be red?
    *(Easy)*

    A.  1/3

    B.  1/8

    C.  3/8

    D.  1/24

**Answer: B. 1/8**
The total number of marbles is 24, and the number of red marbles is 3. Thus the probability of picking a red marble from the jar is 3/24=1/8.


**MATHEMATICS**

41. A die is rolled several times. What is the probability that a 3 will not appear before the third roll of the die?
    *(Rigorous)*

    A.   1/3

    B.   25/216

    C.   25/36

    D.   1/216

**Answer: C. 25/36**

The probability that a 3 will not appear before the third roll is the same as the probability that the first two rolls will consist of numbers other than 3. Since the probability of any one roll resulting in a number other than 3 is 5/6, the probability of the first two rolls resulting in a number other than 3 is (5/6) x (5/6) = 25/36.

**MATHEMATICS**

42. A baseball team has a 60% chance of winning any particular game in a seven-game series. What is the probability that it will win the series by winning games six and seven?
    *(Rigorous)*

    A.   8.3%

    B.   36%

    C.   50%

    D.   60%

**Answer: A. 8.3%**

The team needs to win four games to win a seven-game series. Thus, if the team needs to win games six and seven to win the series, it can only win two of the first five games. There are 10 different ways in which the team can win two of the first five games (where W represents a win and L represents a loss): WWLLL, WLLLW, WLLWL, WLWLL, LWWLL, LWLWL, LWLLW, LLWLW, LLWWL, and LLLWW. This is also the same as the number of combinations of 5 taken 2 at a time. In each case, the probability that the team will win two games and lose three is the following:

$$P(2 \text{ wins and 3 losses}) = (0.6)^2(0.4)^3 = 0.02304$$

Since there are ten ways this can occur, the probability of two wins out of five consecutive games is

$$P(2 \text{ wins out of 5 games}) = 10(0.6)^2(0.4)^3 = 0.2304$$

The probability of the team winning the last two consecutive games is simply $0.6^2$, or 0.36. Then, the probability that the team will win the series by winning games six and seven is the following:

$$P(\text{Series win with wins in games 6 and 7}) = (0.2304)(0.36) = 0.082944$$

43. The scalar multiplication of the number 3 with the matrix $\begin{pmatrix} 2 & 1 \\ 3 & 5 \end{pmatrix}$ yields

    *(Easy)*

    A.  33

    B.  $\begin{pmatrix} 6 & 1 \\ 9 & 5 \end{pmatrix}$

    C.  $\begin{pmatrix} 2 & 3 \\ 3 & 15 \end{pmatrix}$

    D.  $\begin{pmatrix} 6 & 3 \\ 9 & 15 \end{pmatrix}$

**Answer: D.** $\begin{pmatrix} 6 & 3 \\ 9 & 15 \end{pmatrix}$

In scalar multiplication of a matrix by a number, each element of the matrix is multiplied by that number.

**MATHEMATICS**

44. **Evaluate the following matrix product:**

$$\begin{pmatrix} 2 & 1 & 3 \\ 2 & 2 & 4 \end{pmatrix} \begin{pmatrix} 6 & 5 \\ 2 & 1 \\ 2 & 7 \end{pmatrix}$$

*(Rigorous)*

A. $\begin{pmatrix} 20 & 32 & 24 \\ 24 & 40 & 48 \end{pmatrix}$

B. $\begin{pmatrix} 20 & 32 \\ 40 & 24 \\ 24 & 48 \end{pmatrix}$

C. 116

D. $\begin{pmatrix} 20 & 32 \\ 24 & 40 \end{pmatrix}$

**Answer: D.** $\begin{pmatrix} 20 & 32 \\ 24 & 40 \end{pmatrix}$

The product of a 2x3 matrix with a 3x2 matrix is a 2x2 matrix. This alone should be enough to identify the correct answer. Each term in the 2x2 matrix is calculated as described below.

Matrix 1, row 1 multiplied by matrix 2, column 1 yields entry 1, 1: 2x6 + 1x2 + 3x2 = 12 + 2 + 6 = 20.

Matrix 1, row 1 multiplied by matrix 2, column 2 yields entry 1, 2: 2x5 + 1x1 + 3x7 = 10 + 1 + 21 = 32.

Matrix 1, row 2 multiplied by matrix 2, column 1 yields entry 2, 1: 2x6 + 2x2 + 4x2 = 12 + 4 + 8 = 24.

Matrix 1, row 2 multiplied by matrix 2, column 2 yields entry 2, 2: 2x5 + 2x1 + 4x7 = 10 + 2 + 28 = 40.

**MATHEMATICS**

45. Solve the following matrix equation:

$$3x + \begin{pmatrix} 1 & 5 & 2 \\ 0 & 6 & 9 \end{pmatrix} = \begin{pmatrix} 7 & 17 & 5 \\ 3 & 9 & 9 \end{pmatrix}$$

*(Average)*

A. $\begin{pmatrix} 2 & 4 & 1 \\ 1 & 1 & 0 \end{pmatrix}$

B. 2

C. $\begin{pmatrix} 8 & 23 & 7 \\ 3 & 15 & 18 \end{pmatrix}$

D. $\begin{pmatrix} 9 \\ 2 \end{pmatrix}$

**Answer: A.** $\begin{pmatrix} 2 & 4 & 1 \\ 1 & 1 & 0 \end{pmatrix}$

Use the basic rules of algebra and matrices.

$$3x = \begin{pmatrix} 7 & 17 & 5 \\ 3 & 9 & 9 \end{pmatrix} - \begin{pmatrix} 1 & 5 & 2 \\ 0 & 6 & 9 \end{pmatrix} = \begin{pmatrix} 6 & 12 & 3 \\ 3 & 3 & 0 \end{pmatrix}$$

$$x = \frac{1}{3}\begin{pmatrix} 6 & 12 & 3 \\ 3 & 3 & 0 \end{pmatrix} = \begin{pmatrix} 2 & 4 & 1 \\ 1 & 1 & 0 \end{pmatrix}$$

**MATHEMATICS**

46. Find the value of the determinant of the matrix.

$$\begin{vmatrix} 2 & 1 & -1 \\ 4 & -1 & 4 \\ 0 & -3 & 2 \end{vmatrix}$$

*(Average)*

A.  0

B.  23

C.  24

D.  40

**Answer: C. 24**
To find the determinant of a matrix without the use of a graphing calculator, repeat the first two columns as shown:

```
2  1    -1   2    1
4 -1     4   4   -1
0 -3     2   0   -3
```

Starting with the top left-most entry (2), multiply the three numbers in the diagonal going down to the right: $2(-1)(2) = -4$. Do the same starting with 1: $1(4)(0) = 0$. Repeat starting with $-1$: $-1(4)(-3) = 12$. Adding these three numbers yields 8. Repeat the same process starting with the top right-most entry, 1. That is, multiply the three numbers in the diagonal going down to the left: $1(4)(2) = 8$. Do the same starting with 2: $2(4)(-3) = -24$. Repeat starting with $-1$: $-1(-1)(0) = 0$. Add these together to get $-16$. To find the determinant, subtract the second result from the first: $8 - (-16) = 24$.

**MATHEMATICS**

47. Which of the following is a recursive definition of the sequence $\{1, 2, 2, 4, 8, 32, \ldots\}$?

    *(Average)*

    A.   $N_i = 2N_{i-1}$

    B.   $N_i = 2N_{i-2}$

    C.   $N_i = N_{i-1}^2$

    D.   $N_i = N_{i-1}N_{i-2}$

**Answer: D.** $N_i = N_{i-1}N_{i-2}$

Test each answer or look at the pattern of the numbers in the sequence. Note that each number (with the exception of the first two) is the product of the preceding two numbers. In recursive form using index *i*, the expression is $N_i = N_{i-1}N_{i-2}$, and the correct answer is D.

48. What is the sum of the first 20 terms of the geometric sequence:

    (2, 4, 8, 16, 32…)?

    *(Rigorous)*

    A.   2,097,150

    B.   1,048,575

    C.   524,288

    D.   1,048,576

**Answer: A. 2,097,150**

For a geometric sequence $a, ar, ar^2, \ldots, ar^n$, the sum of the first *n* terms is given by $\frac{a(r^n - 1)}{r - 1}$. In this case, $a = 2$ and $r = 2$. Thus, the sum of the first 20 terms of the sequence is $\frac{2(2^{20} - 1)}{2 - 1} = 2{,}097{,}150$.

**MATHEMATICS**

49. Find the sum of the first one hundred terms in the progression.

    (−6, −2, 2…)

    *(Rigorous)*

    A.   19,200

    B.   19,400

    C.   −604

    D.   604

**Answer: A. 19,200**
Examine the pattern of the sequence. To find the 100$^{th}$ term, use the following expression:

$$t_{100} = -6 + 99(4) = 390$$

To find the sum of the first 100 terms, use

$$S = \frac{100}{2}(-6+390) = 19200$$

50. If an initial deposit of $1,000 is made to a savings account with a continuously compounded annual interest rate of 5%, how much money is in the account after 4.5 years?
    *(Average)*

    A.   $1,200.00

    B.   $1,225.00

    C.   $1,245.52

    D.   $1,252.32

**Answer: D. $1,252.32**
Use the formula for continually compounded interest: $A = Pe^{rt}$ where $A$ is the amount in an account with a principal of $P$ and annual interest rate $r$ compounded continually for $t$ years. Then:

$$A = \$1{,}000 e^{0.05 \cdot 4.5} = \$1{,}252.32$$

**MATHEMATICS**

## Answer Key

| | | | |
|---|---|---|---|
| 1. | B | 26. | C |
| 2. | C | 27. | C |
| 3. | B | 28. | A |
| 4. | C | 29. | D |
| 5. | D | 30. | D |
| 6. | B | 31. | B |
| 7. | C | 32. | A |
| 8. | B | 33. | B |
| 9. | A | 34. | A |
| 10. | A | 35. | B |
| 11. | B | 36. | C |
| 12. | C | 37. | B |
| 13. | B | 38. | D |
| 14. | A | 39. | D |
| 15. | C | 40. | B |
| 16. | B | 41. | C |
| 17. | D | 42. | A |
| 18. | D | 43. | D |
| 19. | B | 44. | D |
| 20. | D | 45. | A |
| 21. | B | 46. | C |
| 22. | A | 47. | D |
| 23. | A | 48. | A |
| 24. | C | 49. | A |
| 25. | A | 50. | D |

## Rigor Table

| | Easy 20% | Average 40 % | Rigorous 40% |
|---|---|---|---|
| Questions | 3, 6, 11, 12, 19, 25, 28, 36, 40, 43, | 1, 2, 5, 9, 10, 14, 17, 18, 21, 24, 26, 27, 31, 32, 33, 39, 45, 46, 47, 50 | 4, 7, 8, 13, 15, 16, 20, 22, 23, 29, 30, 34, 35, 37, 38, 41, 42, 44, 48, 49 |

**MATHEMATICS**

www.ingramcontent.com/pod-product-compliance
Lightning Source LLC
LaVergne TN
LVHW061319060426
835507LV00019B/2227

9 781607 872030